MIDTOWN TRASH

RICHARD PANCHYK

AMERICA
THROUGH TIME®
ADDING COLOR TO AMERICAN HISTORY

America Through Time is an imprint of Fonthill Media LLC
www.through-time.com
office@through-time.com

Published by Arcadia Publishing by arrangement with Fonthill Media LLC
For all general information, please contact Arcadia Publishing:
Telephone: 843-853-2070
Fax: 843-853-0044
E-mail: sales@arcadiapublishing.com
For customer service and orders:
Toll-Free 1-888-313-2665

www.arcadiapublishing.com

First published 2019

ISBN 978-1-63499-171-1

Typeset in Gotham
Printed and bound in England

CONTENTS

ACKNOWLEDGMENTS

THANKS to Alan Sutton and Kena Longabaugh at America Through Time for their continued support. And of course, thanks to all those whose trash unwittingly wound up in this book, and to the entire army of trash personnel who work hard to keep Midtown clean.

INTRODUCTION

NEW YORK IS LITERALLY A CITY OF TRASH. A New York City Sanitation Department study found that in 2017, New York City residents threw out 3.1 million tons of trash. New York City sanitation workers collect more than 2,200 tons of garbage and recycling in Manhattan on a daily basis. That's a lot of garbage, and that does not even include the thousands of tons of commercial trash that is hauled away by private carters! Many parts of Manhattan are actually built on garbage, landfill that included everything from pieces of broken ships to bricks and construction debris. Manhattan Island used to be significantly narrower in places, before it was widened with trash. One of the city's most popular attractions, the World's Fair (now Flushing Meadow Corona Park), was constructed on top of a garbage dump. Rikers Island was expanded to double its original size thanks to garbage, and LaGuardia Airport was later built on some of that same trash that was transported from Rikers to Queens.

In this book, we'll take a closer look at Midtown Manhattan's trash—the small and the large, the beautiful and the disgusting, the mundane and the bizarre. The trash in this book is place-specific history, a snapshot of Midtown life both past and present, as seen through garbage. It's a photographic history of the discards of several different categories of Midtown people—its residents, workers, and tourists, all of whom coexist and contribute to the many tons of trash generated in Midtown every single day.

Midtown is an interesting and vibrant location because so many people pass through it on a daily basis, whether they are headed there, or their final destination is elsewhere. Midtown encompasses Manhattan's major transportation hubs—the Port Authority Bus Terminal, Penn Station, and Grand Central Terminal. It also encompasses several of the city's most popular attractions, including the Empire State Building, Madison Square Garden, the New York Public Library and Bryant Park, Rockefeller Center, and St. Patrick's Cathedral, to name just a few.

Midtown Manhattan is a busy place that is continually evolving and changing, and change generates trash. Old buildings are constantly demolished, and new ones are built in their place. On any given day, hundreds of office, commercial, and residential interiors are gutted and rebuilt. Cranes and bulldozers and dumpsters are everywhere. Commuters bustle, residents shop, and tourists explore. There's a lot going on in Midtown all day, every day. All this activity generates a great deal of interesting and varied trash of all kinds.

I've been on a trash trajectory for a while now. I've long been interested in the aesthetic of garbage and where and how people discard it. Several of my previous books have covered abandoned places, and a common theme within those abandoned photographs is trash that has been dumped and strewn about. There was something sadly poetic about the trash, and the abandoned place photographs I took got me thinking: what if I were to take that theme even further and create a book that was an homage to trash? Because pieces of trash are themselves abandoned items, artifacts of our lives.

What makes this topic especially interesting is that trash is ephemeral. This book is unique in that everything you see here is long gone. Most of what I photographed was gone within a matter of a day or two, if not hours or even minutes. These images capture something truly fleeting, in many cases intimate momentary portraits of unwanted and unneeded personal effects. There is something poignant and poetic about trash. It's a circle of life kind of thing. Everything you see in this book was once important, needed, wanted, and utilized. Until it wasn't.

All trash is not created equal. There are a few different categories of trash you'll notice in this book.

1) Intentional trash: Junk that was purposely and properly discarded.
2) Reusable trash: Same as above except it's been thoughtfully exposed and placed so passersby can see and perhaps take and reuse it before the trash pickup. File under "one man's trash is another man's treasure."
3) Intentional litter: The garbage that people left behind because they could not be bothered to throw it away properly. This includes trash left on top of closed garbage cans. Littering in New York City is a crime punishable by a fine of at least $75.
4) Fallen litter: Items that were unintentionally dropped but then left where they fell for someone else to clean. This category most often consists of food or beverage items. It also includes items dropped by young children who may have realized they dropped something but could not express it in words.
5) Lost items: Items that were unwittingly dropped and either remained where they fell or were picked up and placed somewhere prominent in case the owner returned.

The vast majority of these photographs were taken over a four-month period between November 20, 2018, and March 21, 2019, on the streets of Midtown Manhattan. The making of a book such as this one could easily turn into a years-long (and hundreds of pages long) documentary project, but I thought it would be more interesting and tell a better story if I contained it to a short, well-defined time frame. Winter is an interesting season for trash, not just because it contains the Christmas and New Year's holidays, but also because it can get pretty messy with snow, slush, ice, and mud.

I had one rule when taking these photographs: no interference. I did not pose, move, or touch any of the trash. It is all just as I found it, even though sometimes that meant shadows or unflattering angles. Sometimes closeups worked best to show details of the trash, and other times context was more interesting than the details.

I hope you will enjoy this journey into the world of Midtown's trash.

1

Garbage Business

NYC TRASH CAN. Sixth Avenue and 46th Street. The typical Midtown Manhattan trash can, with warning sign attached. In 2019, the NYC Department of Sanitation, in partnership with the Van Alen Institute and the Industrial Designers Society of America/American Institute of Architects, held a competition called Better Bin for the design of the trash can of the future. Almost 200 submissions were received from entrants in six continents. Three finalists will receive $40,000 to develop prototypes to be used in the city.

NYC SANITATION TRUCK. Eighth Avenue between 34th and 35th Streets. The Department of Sanitation uses a fleet of more than 2,100 garbage trucks to collect 12,000 tons of trash every day. Garbage trucks are usually on the move, so it's rare to capture an active one in a stationary moment.

STREET SWEEPER. Eighth Avenue and 39th Street. There are 430 of these mechanical brooms, AKA street sweepers, that clean the 6,000 miles of city streets. This one is traveling north on Eighth Avenue on a rainy December morning.

DON'T LITTER. Eighth Avenue between 34th and 35th Streets. An appropriate slogan to place on the side of a garbage truck, don't you think?

TRASH RECEPTACLES. Broadway and 52nd Street. Aside from ordinary trash cans, there are lots of trios like this around Midtown—one bin is for plastic and glass bottles, one for paper, and one for general trash. Aside from the obvious advantages offered by giving New Yorkers a proper place to recycle, the refuse is hidden from sight, making the streets less ... trashy.

NESTED TRASH CANS. 46th Street between Sixth and Seventh Avenues. Sometimes even trash itself is trash. Is that a double negative? Trash can sometimes pose great philosophical dilemmas.

TRASH CANS. Broadway between 47th and 48th Streets. Times Square Alliance trash cans stacked up in the median on Broadway. The Alliance was founded in 1992 to improve and promote Times Square. They play a big part in keeping the tourist-filled area clean on a day-to-day basis.

TRASH CAN. Fifth Avenue between 45th and 46th Streets. A Grand Central Partnership can. Local neighborhood alliances really do wonders for keeping Midtown streets and sidewalks clean on a daily basis.

TRASH CAN. 46th Street between Madison and Fifth Avenues. This simple yet elegant stone can (can a can be called a can if it is stone? Or can't it?) won't rust and will withstand even the most powerful trash.

THEATER TRASH CAN. 42nd Street between Seventh and Eighth Avenues. A small trash can is wedged at the bottom of the stairs of the New Victory Theater. Originally built in 1900, the theater showed XXX rated films in the 1970s and was rebuilt and reborn as a more family-friendly place in 1995.

TRASH CANS. Fifth Avenue and 48th Street. Do you see the Fifth Avenue Business Improvement District cans on the corners at this busy intersection? Midtown generally has plenty of trash cans, so there's little excuse for littering.

FIT CAMPUS. 27th Street between Seventh and Eighth Avenues. It's a combination of diligent cleaning crews and plentiful trash cans that contribute to a clean-looking environment. On the right side, you can see three trash cans along the edge of the sidewalk on the Fashion Institute of Technology campus.

TRASH CANS. Grand Central Terminal. These cans have been wrapped with an image of the historic Grand Central Terminal, making them perhaps the most aesthetically pleasing trash cans in Midtown, and also camouflaging them somewhat.

GREEN CAN. Greeley Square. Another aesthetically pleasing trash can. Greeley Square garbage, represent!

TRASH BAGS AND CAN. Lexington Avenue and 45th Street. Someone has dropped a few trash bags next to a Grand Central Alliance trash can. The official New York City Sanitation guidelines for residential trash placement are: "Place your items curbside on the curb between 4 PM and midnight the evening before your scheduled pickup."

CLEANING EQUIPMENT. Bryant Park. At any one time during a typical Midtown day, there are probably hundreds of cleaning staff on duty from the various business improvement districts. Midtown's population grows by hundreds of thousands of people during weekdays, so keeping it clean is critically important.

TRIO OF CANS. Bryant Park. I think people are more likely to dispose of their trash properly when the cans are brightly colored and hard to miss.

CLEANING EQUIPMENT. Herald Square. The 34th Street Partnership was launched in 1989. At the time it was one of New York City's first Business Improvement Districts. It covers a thirty-one-block area centered around 34th Street.

TRASH MAN. Greeley Square. A 34th Street Partnership employee on the job, scouring Greeley Square for garbage. These folks do make a difference and help keep tables and chairs in Midtown plazas and parks clean of debris.

CLEANING CREW MEMBER. 42nd Street between Seventh and Eighth Avenues. A Times Square Alliance employee on the hunt for trash in the Theater District after dark.

TRASH CAN. Herald Square. A peek inside a Midtown cleaning crew trash can reveals that practically all of what's left lying around in public spaces is food trash. Clean up after yourselves, people!

34
34
Litter
Stops Here

STORE
Your Home for True
800.696.1350

BROOM. Rockefeller Plaza. The Rockefeller Center area is pretty clean, thanks in part to the broom in this photograph. I do believe that clean places compel people to be neater and dirty places compel people to feel more inclined to litter.

PREVIOUS PAGE:

▲ **SUBWAY TRASH.** 34th Street Station, Eighth Avenue Line. People always seem to be eating or drinking something while waiting for their subway to arrive. Trash cans are a welcome component of every subway station.

▼ **TRASH ON THE MOVE.** Eighth Avenue and 39th Street. With changing lights and pedestrians bustling around everywhere, it's pretty hard to capture cans in motion, but I managed to do it here. This looks like commercial trash, since the guys are not wearing any local alliance uniforms.

2

DUMPSTER DIVE

DUMPSTERS. Sixth Avenue between 37th and 38th Streets. A bunch of empty dumpsters await commercial trash. According to an article in the Commercial Observer ("Trash Talk: Inside the Fight to Reform NYC's Commercial Trash Industry" 7/11/2018) there are more than 200 carting companies that service the city's commercial trash, collecting 12,000 tons every day! Sanitation rules dictate that businesses must have their trash removed by a private carter.

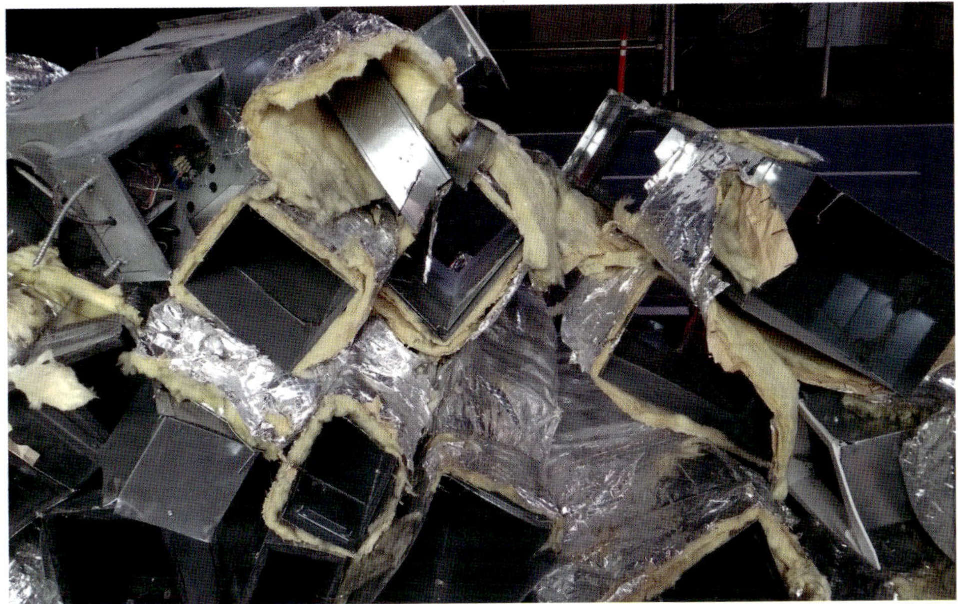

DUCTWORK. 36th Street between Seventh and Eighth Avenues. Demolition trash can be hazardous. I often see guys wearing respiration masks loading dumpsters and it makes me wonder how safe this stuff is to us passersby, especially when the dust is still settling after the junk has been dumped. I'm sure I'm not the only one who has held my breath passing by dumpsters being filled. Incidentally, this is the first photo I took specifically for this book.

SHARP TRASH. Seventh Avenue and 39th Street. Another reminder why New Yorkers need to pay attention as they walk along the city streets and sidewalks! Ouch.

DOOR. Eighth Avenue and 48th Street. A great deal of office and apartment renovation goes on in Midtown, and there is lots of build-out trash generated in the process. From doors to walls to insulation and pipes, it all winds up in dumpsters; some in building loading docks and some out on the street for all to see. Though the sanitation department will remove do-it-yourself oversized residential trash, any residential trash generated by contractors has to be hauled away by private carters.

LIBRARY TRASH. 40th Street between Fifth and Sixth Avenues. Even Midtown's most venerated buildings have an underbelly, a service entrance where trash piles up.

METAL TRASH. 36th Street between Seventh and Eighth Avenues. Dumpster watching is a little like bird watching. It can be fun to see old, familiar trash, but the real thrill is in the rare, harder to identify dumpster garbage.

FLEXIBLE STEEL CONDUIT. 39th Street between Broadway and Sixth Avenue. Dumpster trash often evokes violence because it mainly consists of items that were literally yanked out of buildings.

CONSTRUCTION TRASH. Seventh Avenue and 40th Street. Demolition is not the only generator of large-scale trash. In fact, I've noticed just how much garbage is generated by construction of new buildings. Everything from excess and scrap materials to empty bags and boxes is thrown out on a daily basis at construction sites.

OVERFLOWING DUMPSTER. Fifth Avenue and 40th Street. It's a big city and between residents and workers generates a lot of trash. Even the most photogenic locations in Midtown, such as in front of the public library, are not immune to trash pileups.

▲ **BOXES IN CONTRAST.** 40th Street between Seventh and Eighth Avenues. Skyscrapers may tower high above Midtown's streets, but trash can be towering, too.

▼ **TRASH IN DUMPSTER.** Penn Station 32nd Street Entrance. I guess I never realized just how much trash is literally everywhere. Until I really started looking for it, I didn't truly see it all.

HOMELESS SIGN. 32nd Street between Sixth and Seventh Avenues. An abandoned "help me" sign has been wedged in among other trash in a dumpster. There's something tragically poetic about this.

MAN ON DUMPSTER. 39th Street between Seventh and Eighth Avenues. There is something otherworldly about this image of a guy handling bags of trash atop a dumpster against a bright cloudy sky.

WIRE COILS. 45th Street between Sixth and Seventh Avenues. Broadway shows come and go and with each changing of the dramatic guard, there is plenty of trash generated as sets are deconstructed and reconstructed.

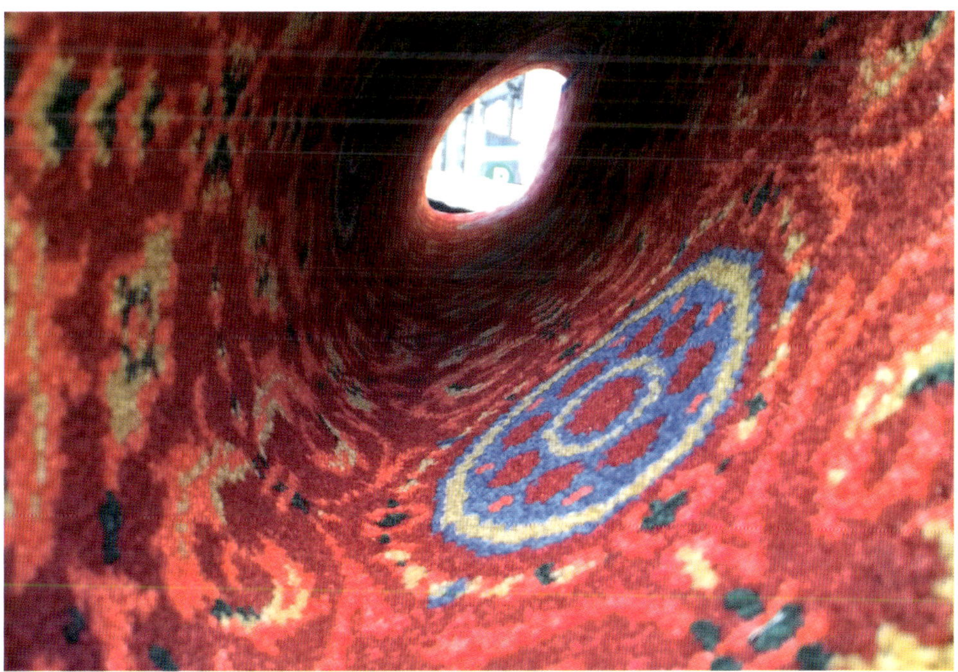

MAGIC CARPET. 43rd Street between Seventh and Eighth Avenues. This was carpet trash from the theater staging the Harry Potter and the Cursed Child Broadway production. As I studied the rug for picture angles, one of the set guys told me to take whatever I wanted.

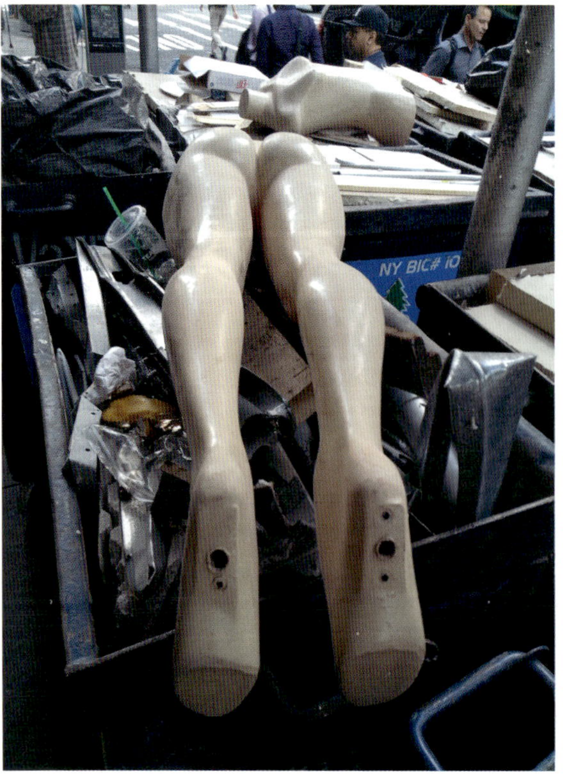

▲ **FABRIC REMNANTS.** 43rd Street between Seventh and Eighth Avenues. More trash from the set of Harry Potter and the Cursed Child.

▼ **DUMMY.** Seventh Avenue at 39th Street. You see some interesting trash in the Fashion District. This photo is from 2017 and predates the book, but I had to take the picture.

DUMPSTER. 45th Street between Seventh and Eighth Avenues. Ever wondered what's inside a typical giant dumpster? Here's a peek at one dumpster's contents. I spy a mattress, an ironing board, a drawer, and random broken pieces of furniture ... along with a pretzel and a soda can, proof that passing New Yorkers view these dumpsters as giant, convenient trash cans, which I guess they are.

MATTRESSES. 33rd Street between Sixth and Seventh Avenues. Trashed mattresses probably from a nearby hotel are stacked up in a dumpster as the Empire State Building looks on.

DUMPSTER WITH FILES. 40th Street between Seventh and Eighth Avenues. Someone's office was cleaned out. Looks like both filing cabinets and their contents were trashed.

RUBBLE IN TRUCK. 46th Street between Fifth and Sixth Avenues. Building renovation generates a lot of rubble. This looks like the remnants of a destroyed wall. The streets of Midtown are crawling with demolition specialty trucks waiting to haul away construction rubble.

SNOWY DUMPSTER. 38th Street between Sixth Avenue and Broadway. I wonder how much of what's in dumpsters was tossed in there by passersby or people from neighboring buildings with random mid-sized crap to discard.

RED CONTAINERS. 39th Street between Broadway and Sixth Avenue. Commercial trash is not usually colorful, so these industrial sized spackle containers immediately caught my eye.

3

Take a Seat

CHAIR WITH LEAF. 45th Street between Eighth and Ninth Avenues. Chairs are a common curbside sight, and they sometimes appear to be in perfect condition.

BROKEN CHAIR. Seventh Avenue between 37th and 38th Streets. On the other hand, some chairs are... well, lost causes.

OFFICE CHAIR. 47th Street between Sixth and Seventh Avenues. This chair seemed like trash and was surrounded with trash. But when I passed by again a day or two later, a homeless person was sitting in it. Was it trash or was it a possession? Maybe a little of both.

SNOWY CHAIR. Eighth Avenue and 37th Street. Proof that everything looks prettier with a coating of snow on it. City snow is quite fleeting, for it seems anything under six inches of snow vanishes from the streets and sidewalks within hours, thanks to salted sidewalks, lots of car and foot traffic, and diligent shoveling.

SUBWAY CHAIR. Eighth Avenue and 35th Street. With trash photos, subject matter and context play an equal role in making for an interesting image. I really liked the aesthetic of this scenario.

LEANING CHAIR. Seventh Avenue and 41st Street. I find that trash chairs usually have personalities. This chair seems inebriated.

UPSIDE DOWN CHAIR. Seventh Avenue and 37th Street. Every time I look at this, I am transfixed by the shiny legs.

NEW AND OLD. Eighth Avenue and 39th Street. Every now and then someone tosses something really old. In this case, a broken piece of late nineteenth-century furniture —perhaps a settee. Trash like this has a story to tell. Too bad I don't speak trash. Well, actually I do. But not nineteenth-century trash, it's a different dialect.

SOFA. 36th between Eighth and Ninth Avenues. This damaged sofa sat here for more than a week before it was finally removed. While most Midtown trash is quite transient, some has staying power, for unknown reasons.

SOFA. 46th between Sixth and Seventh Avenues. When I see big trash like this, I can only imagine how difficult it was to maneuver out of a Midtown building (whether using a freight elevator or the stairs) and to the curb.

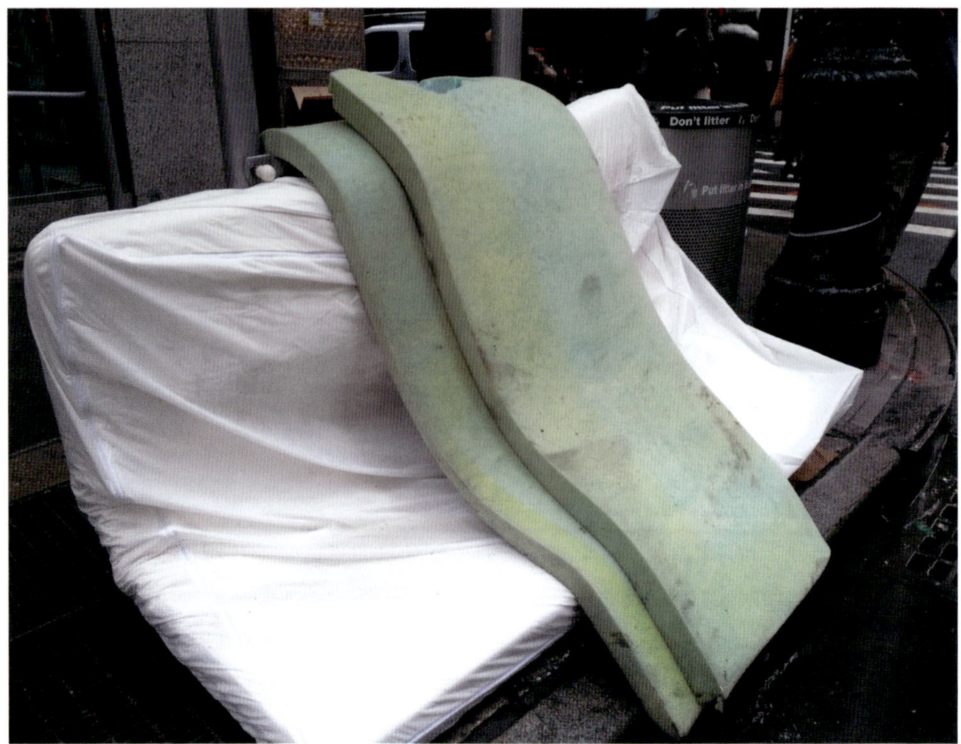

GREEN FOAM. 37th Street and Eighth Avenue. Not exactly a sofa, but in their trash configuration, this mattress and green foam pads look like they would make a quite comfortable place to sit right down.

◄ **BABY CARRIER.** Eighth Avenue and 33rd Street. *Rock-a-bye baby, right in Midtown, all of this traffic's making you frown ... when the light changes, we'll leave this thing here, and carry you home so you don't shed a tear.*

► **STROLLER.** Eighth Avenue between 54th and 55th Streets. This looks like a new stroller with a tag still attached. A broken umbrella is nestled underneath.

4

A Bite to Eat

ORANGE PEEL. 40th Street between Seventh and Eighth Avenues. You can tell a lot about food trash from a quick glance. This orange peel was very recently dropped. It's fresh enough to use in a recipe that calls for orange zest.

▲ **STRAWBERRY.** Bryant Park. Strawberry Fields Forever? No, that's in Central Park. But the sight of a half-buried strawberry did make me wonder if, come spring, the seeds would germinate ... Maybe if everyone buried their fruit remnants in Bryant Park it would transform into an orchard? Free fruit for all!

▼ **FRUIT SPILL.** Seventh Avenue between 35th and 36th Streets. Someone leaving one of the delis along Seventh Avenue with a container of some morning fruit had an accident, probably mere minutes before I took this photo.

PRETZEL. Port Authority Bus Terminal, Eighth Avenue and 41st Street. Putting trash on top of a trash can is like putting money on top of a wallet. I wonder—is it laziness, or is it germaphobia?

WHOLE FOODS BAG. Madison Avenue and 44th Street. Lost or discarded? This brown paper bag was left perched atop a traffic control cabinet. Abandoned paper and plastic bags are a common sight in Midtown.

CANDY BOXES. 38th Street between Broadway and Seventh Avenue. These wholesale-sized candy boxes seem to be overflow from a dumpster heaping with trash.

SUBWAY STATION TRASH. "E" Train 34th Street Station. There are ample trash cans in subway stations, so there's little excuse for leaving garbage on a seat.

ABANDONED LUNCH. Bryant Park. When I came across two unopened containers of salmon rolls on a table in the park, I waited a minute or two to see if their owner had just stepped away briefly, but nobody returned.

GUMMI BEAR. 48th Street between Broadway and Eighth Avenue. Trash is big and trash is small. Sometimes even tiny. And sometimes the tiny trash is the most interesting. The circumstances surrounding this forlorn, fallen gummi candy are pretty easy to imagine. I can see it falling from between the fingers of a toddler in a stroller.

DEAD BAGELS. Sixth Avenue between 46th and 47th Streets. Toasted bagels are quintessentially New York City. Fallen toasted bagels are quintessentially tragic.

FALLEN ROLLS. Eighth Avenue and 36th Street. There is no five-second rule for New York City streets. Once it hits the ground, it's trash. Even freshly baked Kaiser rolls.

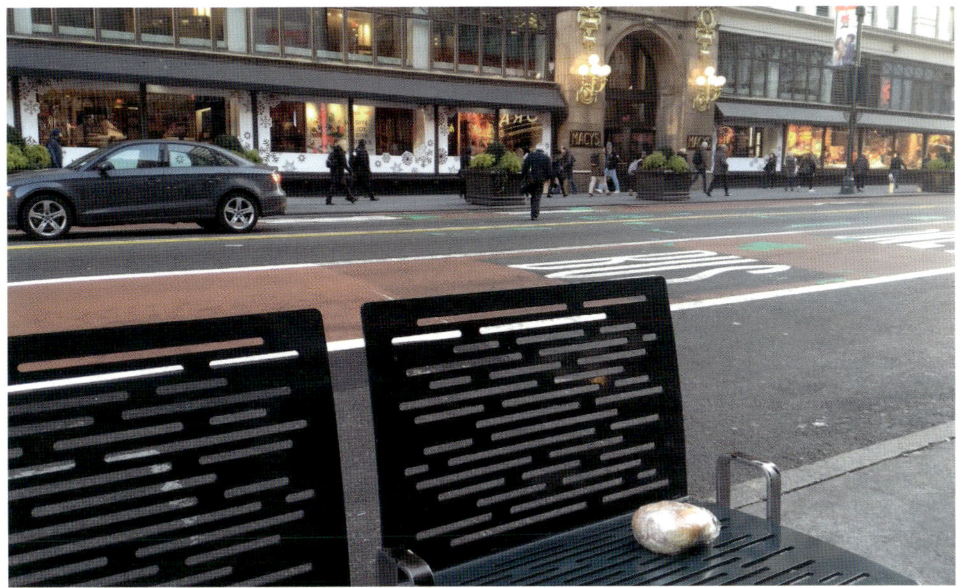

LEFT-BEHIND ROLL. 34th Street between Broadway and Seventh Avenue. A wrapped, buttered Kaiser roll on a bench? Hmmm. No, no, no, it's still trash. Move along.

◀ **FLOWERS AND ICE CREAM.** Seventh Avenue between 35th and 36th Streets. On the street: A vase of wilted flowers next to an untouched Baskin Robbins turkey shaped ice cream cake with sugar cone legs and a caramel praline glaze topping. Yeah, I have no idea.

▶ **LUNCH REMNANTS.** 46th Street between Eighth and Ninth Avenues. A ripped-up piece of junk mail, an empty McNuggets box, and used dipping sauce containers on top of a residential garbage can. The presence of the mail makes me think the culprit was someone who lives here, was finishing up their lunch on the way in, took out the mail from the mailbox, and dumped the refuse atop their trash can. That's the most plausible explanation and even that seems weird.

KETCHUP. 34th Street Subway Station. Taking trash photos in busy Midtown often required waiting until people passed by so I could get a clear shot. Nowhere is this truer than on the steps of a busy subway station.

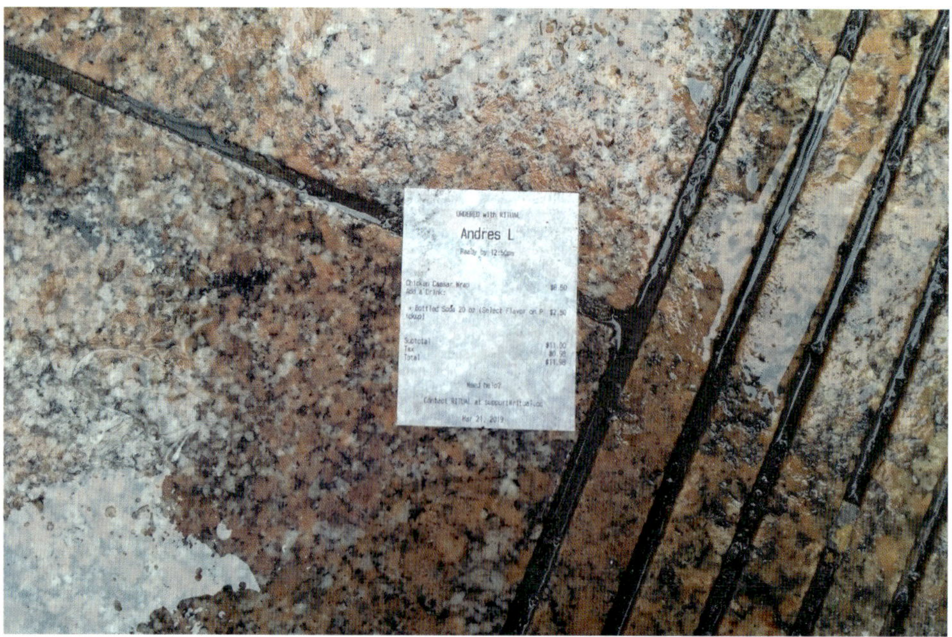

FOOD RECEIPT. Madison Avenue and 45th Street. Looks like Andres had a Caesar salad for lunch today. Every weekday, thousands of office workers hit the streets in search of lunch, and some inevitably drop one remnant or another on their way back to work.

▲ TOMATO BOX. 46th Street between Sixth and Seventh Avenues. With the sheer volume of food deliveries in Midtown, it's amazing that there is not more food mess like this on every block. Whoever was unloading this box from a truck must have dropped it, smushing a bunch of tomatoes in the process. The banana peel is just someone's idea of an artistic touch to add to the aesthetic appeal of the image. Thank you, litterer, for your artistic enhancement.

▼ RECYCLABLE FOOD CONTAINERS. 29th Street between Eighth and Ninth Avenues. According to the New York City government website: "You must place recyclable materials curbside in clear bags or labeled recycling bins between 4 PM and midnight the evening before your scheduled recycling collection day."

WHOLESALE FOOD BOXES. 46th Street between Madison and Fifth Avenues. The restaurant industry generates copious amounts of garbage—both food waste and boxes from all the large-scale supplies that restaurants need to run on a daily basis.

FOOD CART. Fifth Avenue and 45th Street. Every Midtown restaurant generates plenty of trash. Even food carts.

CARDBOARD FOR RECYCLING. Broadway between 39th and 40th Streets. These boxes were put out for recycling in accordance with DSNY rules that state the City accepts "corrugated cardboard boxes (flattened and tied together with sturdy twine)." The list of accepted items is long, but so is the list of prohibited items, including soiled tissues, candy wrappers, cassette tapes, basketballs, light bulbs, and cigarette lighters.

MINI FRIDGES. Eighth Avenue and 43rd Street. According to City rules: "You must make an appointment before discarding any appliance containing Freon or chlorofluorocarbon gas (CFC) from your residence ... You can schedule up to six appliances for CFC removal per appointment. You must place all appliances out with their backs facing the street. Only appliances that are tagged for pick-up will be collected."

5

DRINK UP

PEPSI CAN. 34th Street between Seventh and Eighth Avenues. People will leave trash anywhere they can. But just because it's not on the ground doesn't mean it's not littering!

LITTLE CUP, BIG WHEEL. 45th Street between Sixth and Seventh Avenues. The contrast here struck me, as did the odd color of the liquid in the cup. Ewww.

FOUR CUPS. Port Authority Bus Terminal. The cup placement reminded me of a British Invasion band with the two lead singers in the front and the drummer and keyboardist on an elevated platform behind. Okay, maybe all this trash is getting to me. But you see it, too, right?

CRUSHED CANS. Broadway at 47th Street, 50th Street between Madison and Fifth Avenues. Trash that winds up in the street is inevitably flattened. If I was a famous contemporary artist, I could collect a bunch of these flat cans, make a massive collage, and sell it for $3 million. Maybe $4 million on a good day.

TRACK TRASH. Times Square Station 1-2-3 line. Subway tracks are a common place for litter, and one reason that so many rats can be spotted scurrying along the tracks—unfinished bits of food often wind up down there, making a tasty snack and explaining the hefty size of subway rats. Most commonly seen on tracks are bottles and cans.

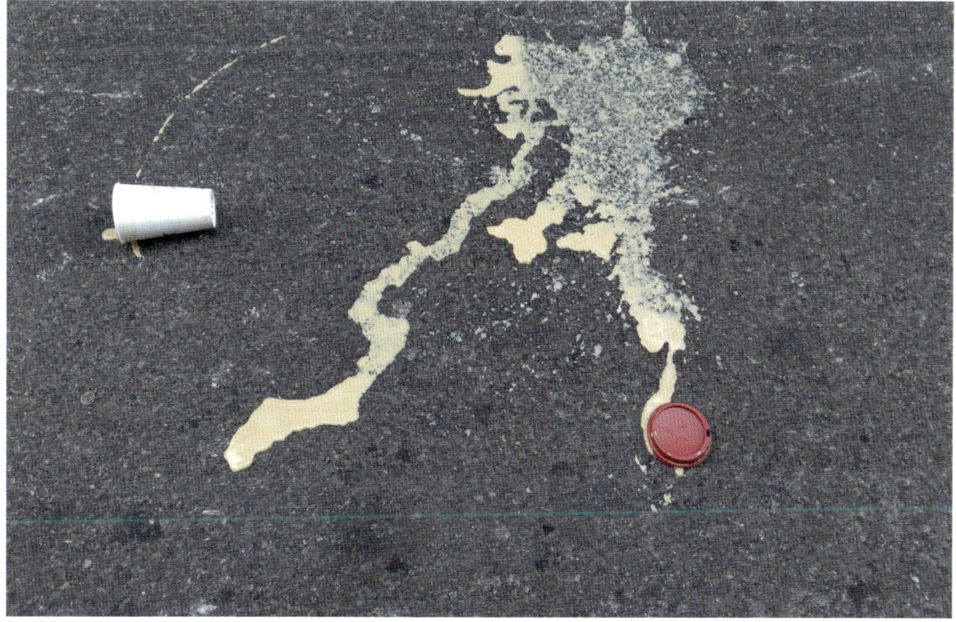

COFFEE SPILL. 43rd Street between Seventh and Eighth Avenues. I call this Crime Scene Trash. The poor cup of coffee met with a violent and unexpected end.

ICE. 43rd Street between Seventh and Eighth Avenues. Ice is the only kind of trash that will vanish completely and quickly. Well, except in winter. On the day this photo was taken the high temperature was 32 degrees.

RED BULL. 45th between Sixth and Seventh Avenues. I do not condone littering, but I must admit I do enjoy litter creative placement.

COMMUTER CUPS. Grand Central Terminal. Though Grand Central is pretty clean, even the most diligent efforts cannot keep such a massive and crowded place spotless every minute of the day. These cups probably weren't here long and were likely going to be gone within minutes of these photographs.

SEVERAL CUPS. 46th Street between Sixth and Seventh Avenues. This looks like several people were standing around sipping their drinks and then decided to move along, without their cups.

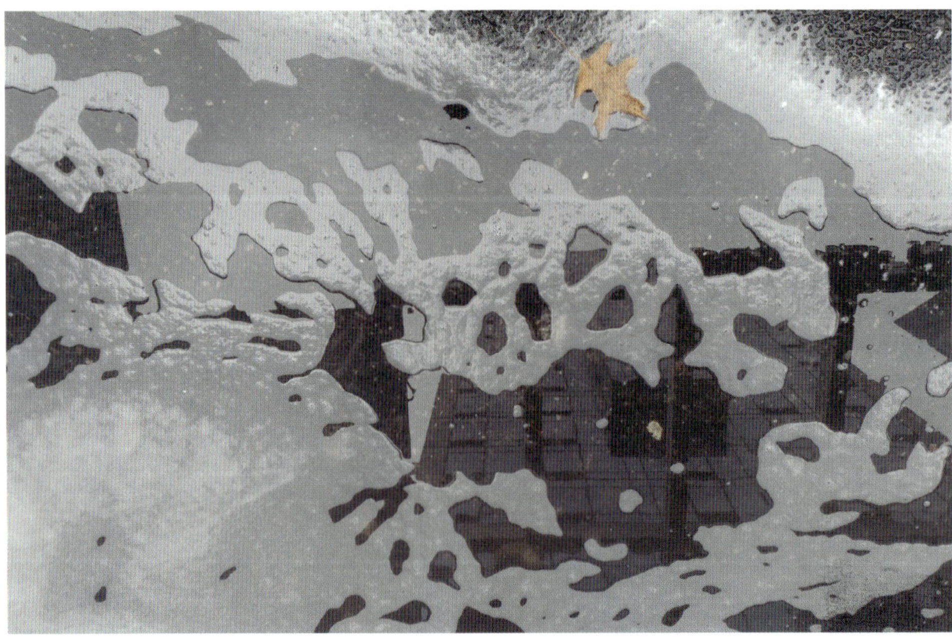

SOAP SUDS. 46th Street between Fifth and Sixth Avenues. Soap suds, spilled coffee, chemicals, and other unknown liquid substances are a type of temporary trash, only present until it drains or evaporates away.

6

TRASHED HOLIDAYS

SANTA BOOK. Eighth Avenue and 35th Street. On January 9, someone decided they didn't want this special 44-page *'Twas the Night Before Christmas* Coca Cola Santa tribute book, but left it out in case I wanted it. Very considerate.

ORNAMENT. Eighth Avenue and 35th Street. A large, cheap, broken Christmas ornament sits in a trash can a week before Christmas. It looks like the cracked eggshell of some rare holiday bird.

HEATER BOX. Eighth Avenue and 30th Street. Typical winter trash. Maybe someone bought this heater locally and didn't want to lug the box on the bus or subway so ditched it here.

CHRISTMAS TREE. 46th Street between Eighth and Ninth Avenues. Come January, it's Christmas tree disposal time, but there are not as many out for the trash as you might think. I would guess that many Midtown residents, whether living in walkups or high-rise apartment complexes, have fake trees because it's easier for city living.

◄ **TREE BRANCH.** Times Square. Timing is everything for this little branch. Either from a discarded tree or from some festive decoration, it serves as a reminder that it's January 9th and Christmas is over. However, the same image taken a few weeks earlier would be a joyful sign of the coming holiday season.

► **TREE BOX.** 40th Street between Eighth and Ninth Avenues. Tree-in-a-box. For an easy urban Christmas. Two reasons this is a rare sighting: 1) Fake trees generally last several years. 2) Most people keep the box to store the tree in until next year.

CONFETTI. Times Square. New Year's Eve in Times Square attracts a huge crowd of a million people, which itself can generate a whole lot of garbage while standing around for hours on end. But the confetti is the real mess; 3,000 pounds of it drop on Times Square every January 1. Though NYC Sanitation cleaning crews do an admirable job, it's impossible to get all the confetti, which has a knack for hiding in crevices and blowing around in the winter wind. These images, and the next four, were all taken on January 2.

WET CONFETTI. Times Square. Wet confetti sticks to all kinds of surfaces.

▲ **CONFETTI AND BOTTLE.** Times Square. A *very* happy New Year!

▼ **GRATE CONFETTI.** Grates like this are great collectors of confetti. Look at that mass of confetti underneath!

FLYING CONFETTI. Times Square. A steady wind blows on January 3, and a few adventurous pieces of confetti attempt to return from whence they came.

LINGERING CONFETTI. Even three weeks after New Year's Day, confetti can still be spotted in corners and nooks around Times Square.

SIGN IN SUBWAY. Times Square Station. This track-trashed sign that probably belonged to a beggar was photographed on January 13 and answers the age-old question: How far into January do you need to say Happy New Year?

GREEN FEATHER. 45th Street between Madison and Vanderbilt Avenues. It's the day after St. Patrick's Day and there are still a few signs left of the festivities. Footnote: A neighborhood cleaning crew member saw me taking this photo and within seconds after I left, he swept up the feather.

GREEN BEADS. 45th between Sixth and Seventh Avenues. Four days after St. Patty's Day and there are still signs of the festivities that took place.

7

Trash Fashion and Accessories

BOOTS. 27th Street between Seventh and Eighth Avenues. Have a boot, leave a boot/need a boot, take a boot? This image was taken near FIT, so hopefully these boots fit.

MORE BOOTS. 46th Street between Madison and Fifth Avenues. A messy, slushy winter storm + a few missteps into a deep, dirty puddle = abandoned boots. I wonder what Footwear Plan B was. Hmm.

SNEAKERS. 32nd Street between Sixth and Seventh Avenues. Free, a pair of semi-fashionable black sneakers that just need a little TLC and they'll be (almost) as good as semi-new!

FANNY PACK. 43nd Street between Sixth Avenue and Broadway. Some items are a tossup between lost or abandoned, but this, sadly, looks like it falls squarely into the lost category. In an area of Midtown crawling with tourists, I bet this kind of thing happens often.

GOGGLES. 45th Street between Fifth and Sixth Avenues. Many trash scenes are pretty ironic (don't you think?). This one's a little too ironic. It's also a reminder of just how messy New York City's streets can get after rain or snowfall.

SINGLE SNEAKER. 30th Street between Eighth and Ninth Avenues. All the lonely sneakers, where do they all come from?

BLACK GLOVE. 45th Street between Fifth and Sixth Avenues. The streets of Midtown in winter are littered with forlorn hats, scarves, and gloves of all kinds and sizes. I should know; I've lost my share over the years.

BLUE GLOVE. 47th Street between Sixth and Seventh Avenues. He crossed the fire line, and sadly, he wasn't fire-retardant. But apparently his glove was.

COMB. Sixth Avenue and 42nd Street. We are so used to trash everywhere that it becomes background noise. Working on this book opened my eyes to little things I'd never have noticed before, like this well-worn comb in the middle of a busy intersection.

HANDBAG. Times Square. Maybe someone stole this, took the wallet and dumped the bag. A handbag just doesn't seem like the kind of item you'd suddenly want to part with on the fly while walking around the city.

PANTS. Eighth Avenue between 29th and 30th Streets. Looks to me like a ripped reject from a homeless person's temporary encampment, probably originally taken from ... the trash. Trashes to trashes, dust to dust.

SWEATSHIRT. 43rd Street between Madison and Fifth Avenues. At least I think it's a sweatshirt, and it's wrapped around a piece of cardboard and laying in front of the Fifth Church of Christ, Scientist.

UMBRELLA. Seventh Avenue between 28th and 29th Streets. You can see destroyed and abandoned umbrellas on any rainy day, not just the really windy ones. In just this one block on a March day I saw three dead umbrellas—two on the ground and one in a trash can.

UMBRELLAS. Eighth Avenue between 37th and 38th Streets/Seventh Avenue and 41st Street. Umbrellas are discarded instantly the moment the wind destroys them. On the spot. Not necessarily in the nearest trash can. Because there is apparently nothing as frustrating as a broken umbrella.

8

Odds and Ends

BAG IN TREE. 34th Street between Seventh and Eighth Avenues. Trash goes where the wind blows. What's the line from the Katy Perry song again? Why yes, I do feel like a plastic bag sometimes. Thank you for asking.

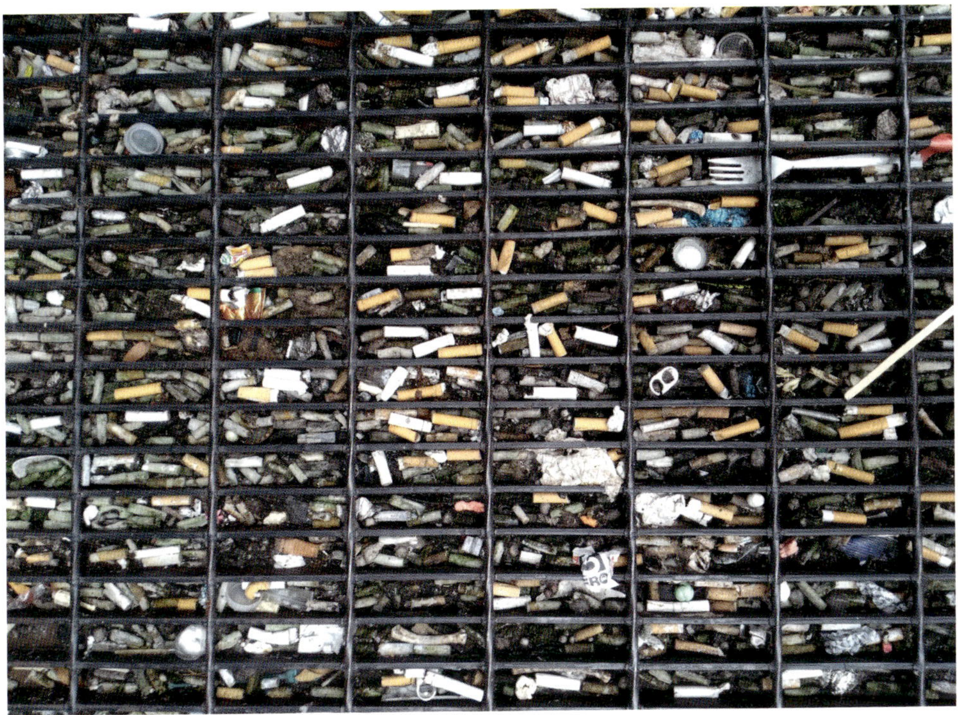

1,000 CIGARETTES. 47th Street between Broadway and Eighth Avenue/Eighth Avenue between 43rd and 44th Streets. Most litter does not lay around forever. Midtown streets and sidewalks get swept regularly, but there are spots that are natural garbage collectors, nooks and crannies that are impossible to clean, such as these grates.

IRON SPIKE. 50th Street between Broadway and Eighth Avenue. This would appear to be an iron train track spike. Is it from the subway system? Metro-North train tracks? And what is it doing on the sidewalk on 50th Street?

INFLATED HAND. Eighth Avenue and 41st Street. The rarely spotted hand balloon not long before it was inevitably flattened by a common non-balloon foot.

STREET SIGN. Times Square. A parking sign waiting for reuse or disposal. I've encountered many pieces of "temporary" trash that litter the sidewalks of Midtown while they await their fate.

CRUTCHES. Broadway between 40th and 41st Streets. I am picturing some kind of Midtown Miracle where a couple of lucky souls were suddenly healed, cast away their now unneeded crutches and sprinted down Broadway singing Hallelujah.

SHATTERED GLASS. 50th Street between Broadway and Eighth Avenue. Just one example of dangerous trash, protruding from a bag in this case, but it's often dumpster trash that features long sharp metal or wood objects. Another reason for pedestrians to mind their surroundings as they walk.

MYSTERY TRASH. Eighth Avenue and 35th Street. Homeless person or trash? In winter, heavily bundled up homeless people can be found on Midtown streets. When they are sleeping, they sometimes resemble a pile of trash.

FALLEN WOOD. Broadway between 35th and 36th Streets. I love capturing newly minted trash ... or temporary trash, assuming it was to be put back in the truck. Questions filled my head as I took in this scene—where is all that wood from, and where is it going? Is it firewood? Is it Norwegian?

SPRING AND FRIENDS. Seventh Avenue between 29th and 30th Streets. There is so much little trash on the streets of Midtown, and much of it is easy to miss. Upon closer inspection, fun mini-trash still-lifes can reveal themselves. I count at least seven distinctly different trash bits in this image.

PARKING TICKET. Seventh Avenue and 28th Street. Parking tickets are usually secured under a windshield wiper, so I'd assume that the recipient of this ticket purposely relegated it to the street.

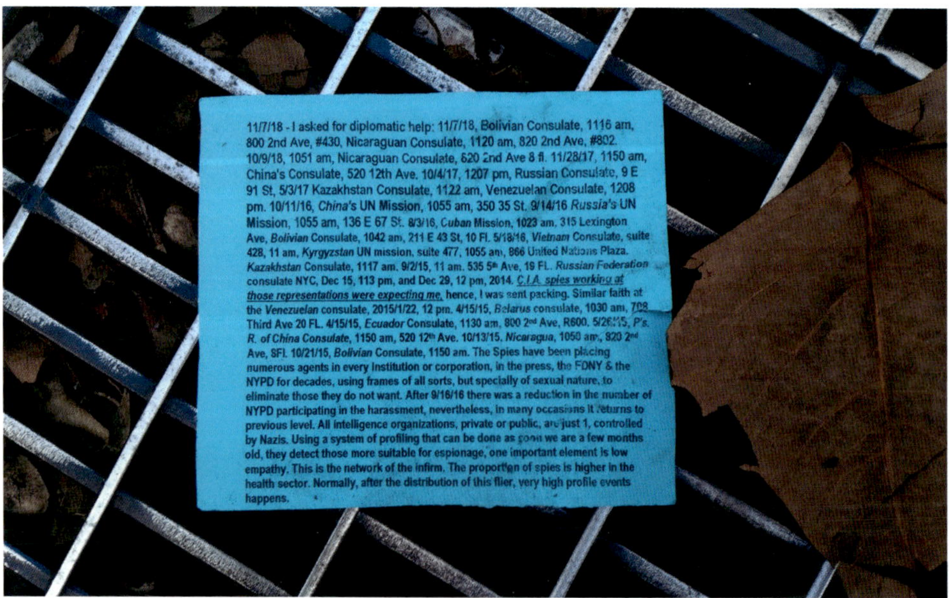

11/7/18 - I asked for diplomatic help: 11/7/18, Bolivian Consulate, 1116 am, 800 2nd Ave, #430, Nicaraguan Consulate, 1120 am, 820 2nd Ave, #802. 10/9/18, 1051 am, Nicaraguan Consulate, 820 2nd Ave 8 fl. 11/28/17, 1150 am, China's Consulate, 520 12th Ave. 10/4/17, 1207 pm, Russian Consulate, 9 E 91 St, 5/3/17 Kazakhstan Consulate, 1122 am, Venezuelan Consulate, 1208 pm. 10/11/16, *China's* UN Mission, 1055 am, 350 35 St. 9/14/16 *Russia's* UN Mission, 1055 am, 136 E 67 St. 8/3/16, *Cuban* Mission, 1023 am. 315 Lexington Ave, *Bolivian* Consulate, 1042 am, 211 E 43 St, 10 Fl. 5/18/16, *Vietnam* Consulate, suite 428, 11 am, *Kyrgyzstan* UN mission, suite 477, 1055 am, 866 United Nations Plaza. *Kazakhstan* Consulate, 1117 am. 9/2/15, 11 am. 535 5th Ave, 19 FL. *Russian Federation* consulate NYC, Dec 15, 113 pm, and Dec 29, 12 pm, 2014. *C.I.A. spies working at those representations were expecting me,* hence, I was sent packing. Similar faith at the *Venezuelan* consulate, 2015/1/22, 12 pm. 4/15/15, *Belarus* consulate, 1030 am, 708 Third Ave 20 FL. 4/15/15, *Ecuador* Consulate, 1130 am, 800 2nd Ave, R600. 5/26/15, *P's. R. of China Consulate*, 1150 am, 520 12th Ave. 10/13/15. *Nicaragua*, 1050 am, 820 2nd Ave, 8Fl. 10/21/15, *Bolivian* Consulate, 1150 am. The Spies have been placing numerous agents in every Institution or corporation, in the press, the FDNY & the NYPD for decades, using frames of all sorts, but specially of sexual nature, to eliminate those they do not want. After 9/16/16 there was a reduction in the number of NYPD participating in the harassment, nevertheless, in many occasions it returns to previous level. All intelligence organizations, private or public, are just 1, controlled by Nazis. Using a system of profiling that can be done as soon we are a few months old, they detect those more suitable for espionage, one important element is low empathy. This is the network of the infirm. The proportion of spies is higher in the health sector. Normally, after the distribution of this flier, very high profile events happens.

MINI MESSAGE. 42nd Street between Fifth and Sixth Avenues. Several of these tiny typed messages lay scattered on the sidewalk. People are constantly handing out all kinds of flyers and papers in Midtown, but this is one of the more interesting.

SCRAP PAPER. Eighth Avenue between 33rd and 34th Streets. People are constantly dropping all kinds of papers—everything from business cards to notes and scrawled numbers that could be the combination to a secret vault full of gold.

SHOW POSTER. Eighth Avenue between 46th and 47th Streets. It's only fitting that Theater District trash includes something Broadway-related.

BATHROOM SINK. 43rd Street between Seventh and Eighth Avenues. In my trash travels I have seen everything but the kitchen sink. Bathroom sink, check. No kitchen sink yet. This many trash bags almost always indicates a large apartment building.

CHECK PROTECTOR. 40th Street between Seventh and Eighth Avenues. When the Drama Bookshop was about to close its doors after decades in business in this location, some interesting trash was put out curbside. This vintage 1960s check protector still had the business card of the company that originally sold it, affixed to the back. A handsome business antique, which someone did later rescue. Not all trash deserves its fate!

▲ **FALLEN MAILBOX.** 36th Street between Eighth and Ninth Avenues. Was this green "relay mailbox" (for temporary mail storage purposes) hit by a car, or intentionally toppled over? Is it trash to be removed or will it be repaired and restored to its place?

▼ **TEXTBOOK.** Times Square. Someone was done with the second edition of *Language and Reality* by Devitt and Sterelny, but was kind enough to leave it laying atop a trash bin in case anyone else wanted to learn about the philosophy of language.

COMPUTER TRASH. 43rd Street between Eighth and Ninth Avenues. With technology constantly improving, unwanted computer-related equipment is to be expected. However, the person who put this stuff out must not realize that as of January 2015, you cannot throw out TVs, monitors, computers, laptops, computer mice, or keyboards in the regular trash.

FAX MACHINE. 36th Street between Eighth and Ninth Avenues. Tech trash is usually interesting and often pretty obsolete. Some of what I've seen belongs in the Smithsonian. Well, maybe more like the Trashsonian.

METROCARDS. Lexington Avenue between 43rd and 44th Streets. There is an add-on fee for purchasing an entirely new MetroCard, to encourage subway users to refill their existing cards. You have two years to trade in expired cards (as the two face down ones in this image are), but not everyone knows or cares.

ORANGE LUGGAGE. 45th Street between Sixth and Seventh Avenues/43rd Street between Fifth and Sixth Avenues. Midtown is filled with tourists and hotels, so it's not uncommon to see luggage out by the curb, whether waiting to be tossed in the trunk of a taxi heading to the airport, or broken and discarded as the one at top appears to be, judging from its off balance left wheel.

LEAF TRASH. 40th Street between Fifth and Sixth Avenues. Though bags of leaves are something you don't see a lot of in Midtown, adjacent to tree-filled Bryant Park is one place you will find leaf refuse once late autumn arrives.

PILE OF TRASH. Park Avenue and 42nd Street. With a backdrop of Grand Central Terminal, a pile of trash waits to be picked up by a NYC Sanitation crew. There is a never-ending cycle of trash in Midtown. No sooner are the sidewalks cleared of trash, new trash is out by the curb. No matter the day of the week, there is trash to be found.

▲ **ORANGE BAGS.** 42nd Street between Fifth and Sixth Avenues. Trash bags or art installation? You decide.

▼ **ELMO.** Seventh Avenue between 39th and 4th Streets. The furry little guy is balanced precariously on the edge of a dumpster. This one is not for tickling, he might fall over.

▲ **PLUSH TOY.** Eighth Avenue between 43rd and 44th Streets. Freshly dropped. Hopefully retrieved soon after I took this photo.

▼ **BROKEN LIGHTS.** 41st Street between Lexington and Third Avenues. Broken fluorescent bulb(s) and other debris on a ledge on the side of a phone booth, where lit-up ads can usually be found.

EMPTY LOTS. 48th Street between Sixth and Seventh Avenues/Fifth Avenue at 43rd Street. Midtown is full of post-demolition empty lots waiting for development. Bricks and other rubble, the last remnants of whatever building used to stand there, are mixed in with random trash.

TRASH IN EMPTY LOT. Fifth Avenue between 46th Street and 47th Streets. A bunch of random junk has been dumped in this empty lot.

HARDWARE. 45th Street between Sixth and Seventh Avenues. The nooks and crevices of Midtown are the most likely to retain small trash like this for the longest time.

ODD JUNK IN EMPTY LOT. 44th Street between Sixth and Seventh Avenues. A stove, a potted plant, and a framed artwork sit in a corner of an empty lot behind a wood fence. Kind of sums up what I love about trash. It's so random!

NECK PILLOW AND CUPS. 50th Street between Broadway and Seventh Avenue. Yeah, I don't understand this one either.

PAINTINGS. 45th Street between Eighth and Ninth Avenues. The stuff we can actually see out for the trash, such as these two psychedelic paintings on wood panels, is just a fraction of the garbage concealed within cans and black bags. I bet there is lots of cool Midtown junk that is discarded, never again to be seen by human eyes.

DRAWERS. Eighth Avenue and 50th Street. Some curbside furniture looks like it deserves a second chance, but these cheap drawers seem to be completely trash-worthy.

BIKE DETOUR. Eighth Avenue and 34th Street. An abandoned and defaced bicycle detour sign on the ground clearly has stories to tell.

OFFICE PAPERS. 40th Street between Seventh and Eighth Avenues. It's rare to see notes and personal business notes and papers out in the open like this; they're usually either shredded or sealed from view in opaque bags. Hey Veronica Burton, whoever and wherever you are, you made the big time now!

ASPHALT. Seventh Avenue between 50th and 51st Streets. Looks like a chunk of asphalt broke through a garbage bag that was being put out for the trash and was left where it fell.

GUITAR PICK. Sixth Avenue between 44th and 45th Streets. I've always had an eagle eye for spotting the little things, going back to my childhood when I'd scour the ground for rocks. There is a Guitar Center just a few blocks from where this was dropped.

BROOM. 43rd Street between Fifth and Sixth Avenues. I was walking along 43rd Street when this well-worn old broom caught my eye. It was laying fallen on the street behind a parked car, as if it had just suddenly collapsed from exhaustion.